MW00681545

HOME is where the CAT IS

BRYAN WOOLLEY

Illustrated by

CHARLES SHAW

BRIGHT SKY PRESS

Home is where the cat is.

He is the **SOUL** of the house.

He moves silently from

ROOM TO ROOM,

up and down the stairs,

IN AND OUT the door.

He inhabits the

WHOLE PLACE at once.

On the patio, he lifts his nose

and **SNIFFS** the air.

The sun **GLISTENS**

on his dense black coat.

His ears move, one and then the other.

When the blue jay in the tree cries his warning,

he hears but **DOES NOT CARE.**

He dashes across the yard and

'round the corner of the house,

RATTLING the dead leaves

that lie on the ground.

He **DASHES BACK** again.

He pauses at the high fence,

measures its distances and angles.

HE LAUNCHES himself

at just the right speed and trajectory.

He sits on the fence in the sun.

He is like a sphinx.

His **TAIL TWITCHES**

at the chatter of the grackles.

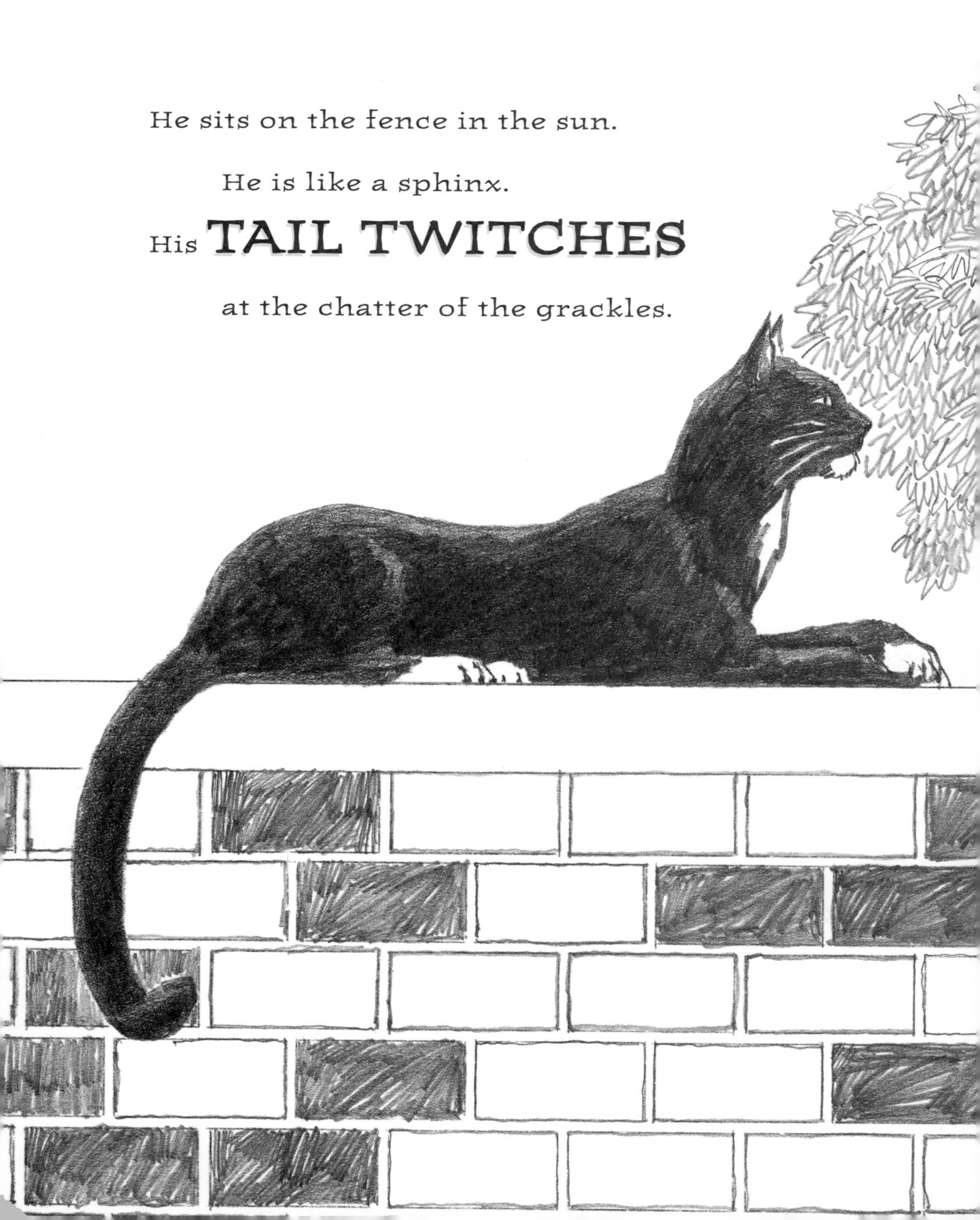

Inside, he knows all

the **BEST PLACES.**

He slides open the closet door

and **SLIPS INSIDE** to nap

in the darkness, in his cave of solitude.

In summer, he **COOLS HIS BELLY**

on the floor tiles.

In winter, he warms himself there,

in the **RECTANGLE OF SUN**

that streams through the window...

or **CURLS HIMSELF**

on the wing chair

in the living room and covers his nose

with the tip of his tail.

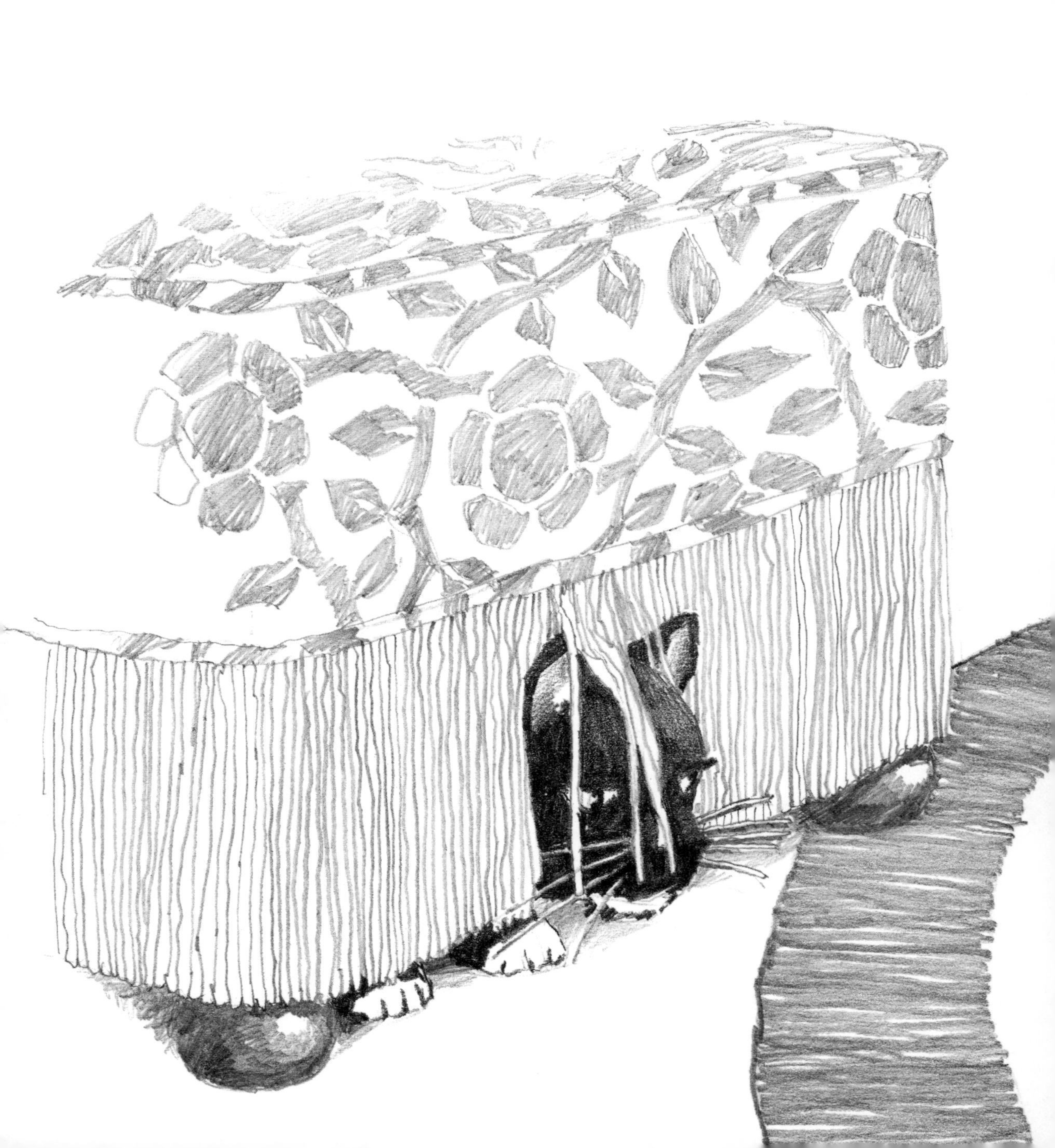

He has **SECRET PLACES**

everywhere. When we call him,

he does not come unless he wants to.

He plays **HIDE-AND-GO-SEEK**

with us. And tag.

He stalks us and grabs our ankles.

Sometimes he plays too rough.

His claws are **SHARP.**

If we yell, he runs to one

of his secret places, but he is not afraid.

He has the **VOICE OF AN ANGEL.**

His name is **JONNY**.

Before him were Hodge and Ace and Pussycat.

The cat has always

OWNED THE PLACE,

but has always shared it with us.

We have always been

GRATEFUL.